The Incredible World of Insects

Incredible Grasshoppers

By Susan Ashley

Gareth Stevens Publishing

Please visit our website, www.garethstevens.com. For a free color catalog of all our high-quality books, call toll free 1-800-542-2595 or fax 1-877-542-2596.

Library of Congress Cataloging-in-Publication Data

Ashley, Susan.
Incredible grasshoppers / Susan Ashley.
p. cm. — (The incredible world of insects)
Includes index.
ISBN 978-1-4339-4588-5 (pbk.)
ISBN 978-1-4339-4589-2 (6-pack)
ISBN 978-1-4339-4587-8 (library binding)
1. Grasshoppers—Juvenile literature. I. Title.
QL508.A2A773 2012
595.7'26—dc22

2010035238

New edition published 2012 by
Gareth Stevens Publishing
111 East 14th Street, Suite 349
New York, NY 10003

Original edition published 2004 by Weekly Reader® Books
An imprint of Gareth Stevens Publishing

Designer: Daniel Hosek
Editors: Mary Ann Hoffman and Kristen Rajczak

Photo credits: Cover, pp. 1, 9, 11, 15 Shutterstock.com; pp. 5, 7, 19 © James P. Rowan; pp. 13, 17 © Robert & Linda Mitchell; p. 21 © Brian Kenney.

Printed in the United States of America

CPSIA compliance information: Batch #CS11GS: For further information contact Gareth Stevens, New York, New York at 1-800-542-2595.

Contents

Boldface words appear in the glossary.

Hiding in the Grass

In summer, watch the grass carefully. You might see **insects** that jump. You might see grasshoppers!

Many grasshoppers are green. This makes it easy for them to hide in the grass and on leaves. This keeps them safe from their **enemies**.

Long and Strong

Grasshoppers have six legs. Their two back legs are long and strong. This helps them jump. A grasshopper can jump 20 times its length!

See Them Fly!

Some grasshoppers don't just jump. They fly, too! These grasshoppers have four wings. They use their two large back wings to fly.

Singing

Male grasshoppers sing to get the attention of female grasshoppers. They do this by rubbing their back legs on their front wings.

Egg to Grasshopper

Grasshoppers **mate** in summer. The female then lays her eggs in a hole in the ground.

The female covers the eggs with sticky matter to keep them safe. The eggs stay in the ground all winter. Most adult grasshoppers don't live through the winter.

A new life **cycle** begins in the spring. The eggs **hatch**. Little grasshoppers come out. These tiny, wingless grasshoppers are called **nymphs**.

The nymphs grow bigger. They **shed** their skin five or six times as their bodies grow. This is called molting. At last, the nymphs become adult grasshoppers.

Glossary

cycle: a group of happenings that take place over and over in the same order

enemy: an animal or person that can cause harm

hatch: to come out of an egg

insect: an animal with six legs, one or two pairs of wings, and three body parts

mate: to come together to make babies

nymph: a young insect that is not fully grown

shed: to get rid of

Index

For More Information

Books

Goldish, Meish. *Leaping Grasshoppers*. New York, NY: Bearport Publishers, 2008.

Nelson, Robin. *Grasshoppers*. Minneapolis, MN: Lerner Publications, 2009.

Web Sites

Grasshopper Information
insected.arizona.edu/ghopperinfo.htm
Learn about grasshoppers as you read interesting facts.

How Grasshoppers Jump
www.st-andrews.ac.uk/~wjh/jumping
Read facts about grasshoppers and see a grasshopper in action.

Publisher's note to educators and parents: Our editors have carefully reviewed these websites to ensure that they are suitable for students. Many websites change frequently, however, and we cannot guarantee that a site's future contents will continue to meet our high standards of quality and educational value. Be advised that students should be closely supervised whenever they access the Internet.